laserpronet.com
Empowering the Laser Workforce

Name: _____

Cell/Text #: _____

What we do at LaserPronet

- Screening Exams
- Professional Development Courses
- Professional Growth Plans
- Certifications

Copyright © 2020

Sukuta Technologies, LLC

All rights reserved

Page

1. Checklist 4
2. Self-Test 36

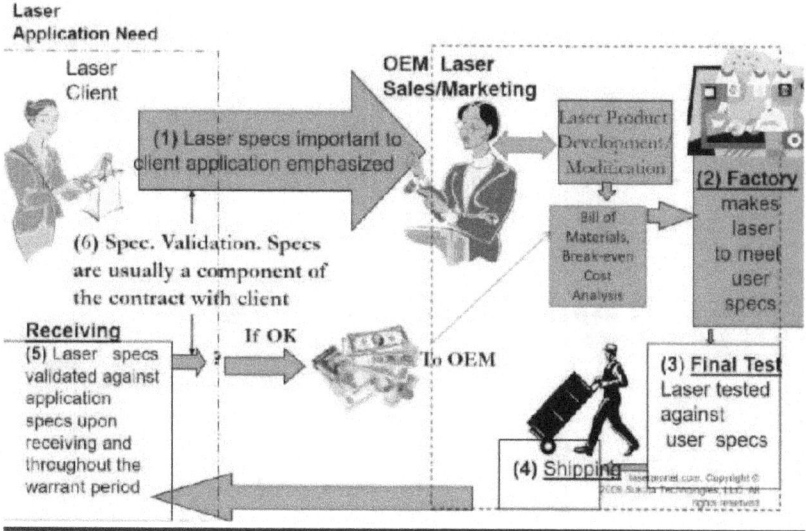

1. Checklist Table of Contents

Date	Project Number	Project Name	Contact

Personal Notes

A. CW Laser Transactional Specifications Checklist

Date	Project Name and #	Client Company		Rep Name:
		Name:		E-mail:
		Url:		Cell/Text #:

Laser Mf._____ Brand Name:_____ Model: _____

A	CW Parameter	Parameter Value (*AD)	Parameter Value (Actual)	PV Diff.	Relative Fluctuation Stability (%) (AD*)	Relative Fluctuation Stability (%) (Actual)	RFS Diff.
1	Power(W)						
2	*Wavelength (nm)*						
3	*Beam Width (mm)*						
4	*Beam Ellipticity*						
5	M^2						
6	*Gaussian Fit (%)*						
7	*BPS($\mu R/°C$)*						
8	*Polarization Ratio*						
C	Tacit Parameter	Parameter Value (*AD)	Parameter Value (Actual)	PV Diff.	Relative Fluctuation Stability (%) (AD*)	Relative Fluctuation Stability (%) (Actual)	RFS Diff.
1	*Threshold*						
2	*Slope Efficiency (%)*						
3	*Warm-up Time (hrs.)*						
4	*Input AC (VAC)*						
5	*Wall-plug Efficiency (%)*						
6	*Useful Lifetime (hrs.)*						
7	*Footprint (m^2)*						

*AD=Application Defined, **BPS=Beam Pointing Stability

B. Pulse Laser Checklist

Date	Project Name and #	Client Company		Rep Name:
		Name:		E-mail:
		Url:		Cell/Text #:

Laser Mf._____Brand Name:_____Model:_____

B	Pulse Parameter	Parameter Value (*AD)	Parameter Value (Actual)	PV Diff.	Relative Fluctuation Stability (%) (AD*)	Relative Fluctuation Stability (%) (Actual)	RFS Diff.
1	Average Power (W)						
2	Energy Per Pulse (mJ/Pulse)						
3	Pulse Duration						
4	Duty Cycle						
5	Rep Rate (kHz)						
6	Peak Power (W)						
C	Tacit Parameter	Parameter Value (*AD)	Parameter Value (Actual)	PV Diff.	Relative Fluctuation Stability (%) (AD*)	Relative Fluctuation Stability (%) (Actual)	RFS Diff.
1	Threshold						
2	Slope Efficiency (%)						
3	Warm-up Time (hrs.)						
4	Input AC (VAC)						
5	Wall-plug Efficiency (%)						
6	Useful Lifetime (hrs.)						
7	Footprint (m^2)						

*AD=Application Defined, **BPS=Beam Pointing Stability

A. CW Laser Checklist

Date	Project Name and #	Client Company		Rep Name:
		Name:		E-mail:
		Url:		Cell/Text #:

Laser Mf._____ Brand Name:_____ Model: _____

A	CW Parameter	Parameter Value (*AD)	Parameter Value (Actual)	PV Diff.	Relative Fluctuation Stability (%) (AD*)	Relative Fluctuation Stability (%) (Actual)	RFS Diff.
1	Power(W)						
2	Wavelength (nm)						
3	Beam Width (mm)						
4	Beam Ellipticity						
5	M^2						
6	Gaussian Fit (%)						
7	BPS($\mu R/^oC$)						
8	Polarization Ratio						
C	Tacit Parameter	Parameter Value (*AD)	Parameter Value (Actual)	PV Diff.	Relative Fluctuation Stability (%) (AD*)	Relative Fluctuation Stability (%) (Actual)	RFS Diff.
1	Threshold						
2	Slope Efficiency (%)						
3	Warm-up Time (hrs.)						
4	Input AC (VAC)						
5	Wall-plug Efficiency (%)						
6	Useful Lifetime (hrs.)						
7	Footprint (m^2)						

*AD=Application Defined, **BPS=Beam Pointing Stability

B. Pulse Laser Checklist

Date	Project Name and #	Client Company	Rep Name:
		Name:	E-mail:
		Url:	Cell/Text #:

Laser Mfr._____Brand Name:_____Model: _____

B	Pulse Parameter	Parameter Value (*AD)	Parameter Value (Actual)	PV Diff.	Relative Fluctuation Stability (%) (AD*)	Relative Fluctuation Stability (%) (Actual)	RFS Diff.
1	Average Power (W)						
2	Energy Per Pulse (mJ/Pulse)						
3	Pulse Duration						
4	Duty Cycle						
5	Rep Rate (kHz)						
6	Peak Power (W)						
C	**Tacit Parameter**	Parameter Value (*AD)	Parameter Value (Actual)	PV Diff.	Relative Fluctuation Stability (%) (AD*)	Relative Fluctuation Stability (%) (Actual)	RFS Diff.
1	Threshold						
2	Slope Efficiency (%)						
3	Warm-up Time (hrs.)						
4	Input AC (VAC)						
5	Wall-plug Efficiency (%)						
6	Useful Lifetime (hrs.)						
7	Footprint (m^2)						

*AD=Application Defined, **BPS=Beam Pointing Stability*

Personal Notes

Personal Notes

A. CW Laser Checklist

Date	Project Name and #	Client Company	Rep Name:
		Name:	E-mail:
		Url:	Cell/Text #:

Laser Mfr._____ Brand Name:_____ Model:_____

A	CW Parameter	Parameter Value (*AD)	Parameter Value (Actual)	PV Diff.	Relative Fluctuation Stability (%) (AD*)	Relative Fluctuation Stability (%) (Actual)	RFS Diff.
1	Power(W)						
2	*Wavelength (nm)*						
3	*Beam Width (mm)*						
4	*Beam Ellipticity*						
5	M^2						
6	*Gaussian Fit (%)*						
7	*BPS($\mu R/°C$)*						
8	*Polarization Ratio*						
C	**Tacit Parameter**	Parameter Value (*AD)	Parameter Value (Actual)	PV Diff.	Relative Fluctuation Stability (%) (AD*)	Relative Fluctuation Stability (%) (Actual)	RFS Diff.
1	*Threshold*						
2	*Slope Efficiency (%)*						
3	*Warm-up Time (hrs.)*						
4	*Input AC (VAC)*						
5	*Wall-plug Efficiency (%)*						
6	*Useful Lifetime (hrs.)*						
7	*Footprint (m^2)*						

*AD=Application Defined, **BPS=Beam Pointing Stability

B. Pulse Laser Checklist

Date	Project Name and #	Client Company		Rep Name:
		Name:		E-mail:
		Url:		Cell/Text #:

Laser Mf._____Brand Name:_____Model:_____

B	Pulse Parameter	Parameter Value (*AD)	Parameter Value (Actual)	PV Diff.	Relative Fluctuation Stability (%) (AD*)	Relative Fluctuation Stability (%) (Actual)	RFS Diff.
1	Average Power (W)						
2	Energy Per Pulse (mJ/Pulse)						
3	Pulse Duration						
4	Duty Cycle						
5	Rep Rate (kHz)						
6	Peak Power (W)						
C	Tacit Parameter	Parameter Value (*AD)	Parameter Value (Actual)	PV Diff.	Relative Fluctuation Stability (%) (AD*)	Relative Fluctuation Stability (%) (Actual)	RFS Diff.
1	Threshold						
2	Slope Efficiency (%)						
3	Warm-up Time (hrs.)						
4	Input AC (VAC)						
5	Wall-plug Efficiency (%)						
6	Useful Lifetime (hrs.)						
7	Footprint (m^2)						

*AD=Application Defined, **BPS=Beam Pointing Stability

A. CW Laser Checklist

Date	Project Name and #	Client Company		Rep Name:
		Name:		E-mail:
		Url:		Cell/Text #:

Laser Mf._____ Brand Name:_____ Model:_____

A	CW Parameter	Parameter Value (*AD)	Parameter Value (Actual)	PV Diff.	Relative Fluctuation Stability (%) (AD*)	Relative Fluctuation Stability (%) (Actual)	RFS Diff.
1	Power(W)						
2	Wavelength (nm)						
3	Beam Width (mm)						
4	Beam Ellipticity						
5	M^2						
6	Gaussian Fit (%)						
7	BPS($\mu R/^oC$)						
8	Polarization Ratio						
C	Tacit Parameter	Parameter Value (*AD)	Parameter Value (Actual)	PV Diff.	Relative Fluctuation Stability (%) (AD*)	Relative Fluctuation Stability (%) (Actual)	RFS Diff.
1	Threshold						
2	Slope Efficiency (%)						
3	Warm-up Time (hrs.)						
4	Input AC (VAC)						
5	Wall-plug Efficiency (%)						
6	Useful Lifetime (hrs.)						
7	Footprint (m^2)						

*AD=Application Defined, **BPS=Beam Pointing Stability

B. Pulse Laser Checklist

Date	Project Name and #	Client Company		Rep Name:
		Name:		E-mail:
		Url:		Cell/Text #:

Laser Mf._____Brand Name:_____Model:_____

B	Pulse Parameter	Parameter Value (*AD)	Parameter Value (Actual)	PV Diff.	Relative Fluctuation Stability (%) (AD*)	Relative Fluctuation Stability (%) (Actual)	RFS Diff.
1	Average Power (W)						
2	Energy Per Pulse (mJ/Pulse)						
3	Pulse Duration						
4	Duty Cycle						
5	Rep Rate (kHz)						
6	Peak Power (W)						
C	Tacit Parameter	Parameter Value (*AD)	Parameter Value (Actual)	PV Diff.	Relative Fluctuation Stability (%) (AD*)	Relative Fluctuation Stability (%) (Actual)	RFS Diff.
1	Threshold						
2	Slope Efficiency (%)						
3	Warm-up Time (hrs.)						
4	Input AC (VAC)						
5	Wall-plug Efficiency (%)						
6	Useful Lifetime (hrs.)						
7	Footprint (m^2)						

*AD=Application Defined, **BPS=Beam Pointing Stability

Personal Notes

Personal Notes

A. CW Laser Checklist

Date	Project Name	Client Company	Rep Name:
		Name:	E-mail:
		Url:	Cell/Text #:

Laser Mf._____ Brand Name:_____ Model: _____

A	CW Parameter	Parameter Value (*AD)	Parameter Value (Actual)	PV Diff.	Relative Fluctuation Stability (%) (AD*)	Relative Fluctuation Stability (%) (Actual)	RFS Diff.
1	Power(W)						
2	*Wavelength (nm)*						
3	*Beam Width (mm)*						
4	*Beam Ellipticity*						
5	M^2						
6	*Gaussian Fit (%)*						
7	*BPS(μR/°C)*						
8	*Polarization Ratio*						
C	**Tacit Parameter**	Parameter Value (*AD)	Parameter Value (Actual)	PV Diff.	Relative Fluctuation Stability (%) (AD*)	Relative Fluctuation Stability (%) (Actual)	RFS Diff.
1	*Threshold*						
2	*Slope Efficiency (%)*						
3	*Warm-up Time (hrs.)*						
4	*Input AC (VAC)*						
5	*Wall-plug Efficiency (%)*						
6	*Useful Lifetime (hrs.)*						
7	*Footprint (m^2)*						

*AD=Application Defined, **BPS=Beam Pointing Stability

B. Pulse Laser Checklist

Date	Project Name and #	Client Company	Rep Name:
		Name:	E-mail:
		Url:	Cell/Text #:

Laser Mf._____ Brand Name:_____ Model:_____

B	Pulse Parameter	Parameter Value (*AD)	Parameter Value (Actual)	PV Diff.	Relative Fluctuation Stability (%) (AD*)	Relative Fluctuation Stability (%) (Actual)	RFS Diff.
1	Average Power (W)						
2	Energy Per Pulse (mJ/Pulse)						
3	Pulse Duration						
4	Duty Cycle						
5	Rep Rate (kHz)						
6	Peak Power (W)						
C	Tacit Parameter	Parameter Value (*AD)	Parameter Value (Actual)	PV Diff.	Relative Fluctuation Stability (%) (AD*)	Relative Fluctuation Stability (%) (Actual)	RFS Diff.
1	Threshold						
2	Slope Efficiency (%)						
3	Warm-up Time (hrs.)						
4	Input AC (VAC)						
5	Wall-plug Efficiency (%)						
6	Useful Lifetime (hrs.)						
7	Footprint (m^2)						

*AD=Application Defined, **BPS=Beam Pointing Stability

A. CW Laser Checklist

Date	Project Name	Client Company	Rep Name:
		Name:	E-mail:
		Url:	Cell/Text #:

Laser Mf._____ Brand Name:_____ Model: _____

A	CW Parameter	Parameter Value (*AD)	Parameter Value (Actual)	PV Diff.	Relative Fluctuation Stability (%) (AD*)	Relative Fluctuation Stability (%) (Actual)	RFS Diff.
1	Power(W)						
2	*Wavelength (nm)*						
3	*Beam Width (mm)*						
4	*Beam Ellipticity*						
5	M^2						
6	*Gaussian Fit (%)*						
7	*BPS($\mu R/^\circ C$)*						
8	*Polarization Ratio*						
C	Tacit Parameter	Parameter Value (*AD)	Parameter Value (Actual)	PV Diff.	Relative Fluctuation Stability (%) (AD*)	Relative Fluctuation Stability (%) (Actual)	RFS Diff.
1	*Threshold*						
2	*Slope Efficiency (%)*						
3	*Warm-up Time (hrs.)*						
4	*Input AC (VAC)*						
5	*Wall-plug Efficiency (%)*						
6	*Useful Lifetime (hrs.)*						
7	*Footprint (m^2)*						

*AD=Application Defined, **BPS=Beam Pointing Stability

B. Pulse Laser Checklist

Date	Project Name and #	Client Company	Rep Name:
		Name:	E-mail:
		Url:	Cell/Text #:

Laser Mf._____Brand Name:_____Model:

B	Pulse Parameter	Parameter Value (*AD)	Parameter Value (Actual)	PV Diff.	Relative Fluctuation Stability (%) (AD*)	Relative Fluctuation Stability (%) (Actual)	RFS Diff.
1	Average Power (W)						
2	Energy Per Pulse (mJ/Pulse)						
3	Pulse Duration						
4	Duty Cycle						
5	Rep Rate (kHz)						
6	Peak Power (W)						
C	Tacit Parameter	Parameter Value (*AD)	Parameter Value (Actual)	PV Diff.	Relative Fluctuation Stability (%) (AD*)	Relative Fluctuation Stability (%) (Actual)	RFS Diff.
1	Threshold						
2	Slope Efficiency (%)						
3	Warm-up Time (hrs.)						
4	Input AC (VAC)						
5	Wall-plug Efficiency (%)						
6	Useful Lifetime (hrs.)						
7	Footprint (m^2)						

*AD=Application Defined, **BPS=Beam Pointing Stability

Personal Notes

Personal Notes

A. CW Laser Checklist

Date	Project Name and #	Client Company		Rep Name:
		Name:		E-mail:
		Url:		Cell/Text #:

Laser Mf._____ Brand Name:_____ Model:_____

A	CW Parameter	Parameter Value (*AD)	Parameter Value (Actual)	PV Diff.	Relative Fluctuation Stability (%) (AD*)	Relative Fluctuation Stability (%) (Actual)	RFS Diff.
1	Power(W)						
2	*Wavelength (nm)*						
3	*Beam Width (mm)*						
4	*Beam Ellipticity*						
5	M^2						
6	*Gaussian Fit (%)*						
7	*BPS($\mu R/^\circ C$)*						
8	*Polarization Ratio*						
C	Tacit Parameter	Parameter Value (*AD)	Parameter Value (Actual)	PV Diff.	Relative Fluctuation Stability (%) (AD*)	Relative Fluctuation Stability (%) (Actual)	RFS Diff.
1	*Threshold*						
2	*Slope Efficiency (%)*						
3	*Warm-up Time (hrs.)*						
4	*Input AC (VAC)*						
5	*Wall-plug Efficiency (%)*						
6	*Useful Lifetime (hrs.)*						
7	*Footprint (m^2)*						

*AD=Application Defined, **BPS=Beam Pointing Stability

B. Pulse Laser Checklist

Date	Project Name and #	Client Company	Rep Name:
		Name:	E-mail:
		Url:	Cell/Text #:

Laser Mf._____Brand Name:_____Model:_____

B	Pulse Parameter	Parameter Value (*AD)	Parameter Value (Actual)	PV Diff.	Relative Fluctuation Stability (%) (AD*)	Relative Fluctuation Stability (%) (Actual)	RFS Diff.
1	Average Power (W)						
2	Energy Per Pulse (mJ/Pulse)						
3	Pulse Duration						
4	Duty Cycle						
5	Rep Rate (kHz)						
6	Peak Power (W)						
C	Tacit Parameter	Parameter Value (*AD)	Parameter Value (Actual)	PV Diff.	Relative Fluctuation Stability (%) (AD*)	Relative Fluctuation Stability (%) (Actual)	RFS Diff.
1	Threshold						
2	Slope Efficiency (%)						
3	Warm-up Time (hrs.)						
4	Input AC (VAC)						
5	Wall-plug Efficiency (%)						
6	Useful Lifetime (hrs.)						
7	Footprint (m^2)						

*AD=Application Defined, **BPS=Beam Pointing Stability

A. CW Laser Checklist

Date	Project Name and #	Client Company		Rep Name:
		Name:		E-mail:
		Url:		Cell/Text #:

Laser Mf._____Brand Name:_____Model:_____

A	CW Parameter	Parameter Value (*AD)	Parameter Value (Actual)	PV Diff.	Relative Fluctuation Stability (%) (AD*)	Relative Fluctuation Stability (%) (Actual)	RFS Diff.
1	Power(W)						
2	*Wavelength (nm)*						
3	*Beam Width (mm)*						
4	*Beam Ellipticity*						
5	M^2						
6	*Gaussian Fit (%)*						
7	*BPS(µR/°C)*						
8	*Polarization Ratio*						
C	Tacit Parameter	Parameter Value (*AD)	Parameter Value (Actual)	PV Diff.	Relative Fluctuation Stability (%) (AD*)	Relative Fluctuation Stability (%) (Actual)	RFS Diff.
1	*Threshold*						
2	*Slope Efficiency (%)*						
3	*Warm-up Time (hrs.)*						
4	*Input AC (VAC)*						
5	*Wall-plug Efficiency (%)*						
6	*Useful Lifetime (hrs.)*						
7	*Footprint (m^2)*						

*AD=Application Defined, **BPS=Beam Pointing Stability

B. Pulse Laser Checklist

Date	Project Name and #	Client Company		Rep Name:	
		Name:		E-mail:	
		Url:		Cell/Text #:	

Laser Mf._____ Brand Name:_____ Model: _____

B	Pulse Parameter	Parameter Value (*AD)	Parameter Value (Actual)	PV Diff.	Relative Fluctuation Stability (%) (AD*)	Relative Fluctuation Stability (%) (Actual)	RFS Diff.
1	Average Power (W)						
2	Energy Per Pulse (mJ/Pulse)						
3	Pulse Duration						
4	Duty Cycle						
5	Rep Rate (kHz)						
6	Peak Power (W)						
C	Tacit Parameter	Parameter Value (*AD)	Parameter Value (Actual)	PV Diff.	Relative Fluctuation Stability (%) (AD*)	Relative Fluctuation Stability (%) (Actual)	RFS Diff.
1	Threshold						
2	Slope Efficiency (%)						
3	Warm-up Time (hrs.)						
4	Input AC (VAC)						
5	Wall-plug Efficiency (%)						
6	Useful Lifetime (hrs.)						
7	Footprint (m^2)						

*AD=Application Defined, **BPS=Beam Pointing Stability

Personal Notes

Personal Notes

A. CW Laser Checklist

Date	Project Name and #	Client Company		Rep Name:	
		Name:		E-mail:	
		Url:		Cell/Text #:	

Laser Mf._____Brand Name:_____Model: _____

A	CW Parameter	Parameter Value (*AD)	Parameter Value (Actual)	PV Diff.	Relative Fluctuation Stability (%) (AD*)	Relative Fluctuation Stability (%) (Actual)	RFS Diff.
1	Power(W)						
2	*Wavelength (nm)*						
3	*Beam Width (mm)*						
4	*Beam Ellipticity*						
5	M^2						
6	*Gaussian Fit (%)*						
7	*BPS($\mu R/^{\circ}C$)*						
8	*Polarization Ratio*						
C	**Tacit Parameter**	Parameter Value (*AD)	Parameter Value (Actual)	PV Diff.	Relative Fluctuation Stability (%) (AD*)	Relative Fluctuation Stability (%) (Actual)	RFS Diff.
1	*Threshold*						
2	*Slope Efficiency (%)*						
3	*Warm-up Time (hrs.)*						
4	*Input AC (VAC)*						
5	*Wall-plug Efficiency (%)*						
6	*Useful Lifetime (hrs.)*						
7	*Footprint (m^2)*						

*AD=Application Defined, **BPS=Beam Pointing Stability

B. Pulse Laser Checklist

Date	Project Name and #	Client Company		Rep Name:
		Name:		E-mail:
		Url:		Cell/Text #:

Laser Mf._____ Brand Name:_____ Model:_____

B	Pulse Parameter	Parameter Value (*AD)	Parameter Value (Actual)	PV Diff.	Relative Fluctuation Stability (%) (AD*)	Relative Fluctuation Stability (%) (Actual)	RFS Diff.
1	Average Power (W)						
2	Energy Per Pulse (mJ/Pulse)						
3	Pulse Duration						
4	Duty Cycle						
5	Rep Rate (kHz)						
6	Peak Power (W)						
C	Tacit Parameter	Parameter Value (*AD)	Parameter Value (Actual)	PV Diff.	Relative Fluctuation Stability (%) (AD*)	Relative Fluctuation Stability (%) (Actual)	RFS Diff.
1	Threshold						
2	Slope Efficiency (%)						
3	Warm-up Time (hrs.)						
4	Input AC (VAC)						
5	Wall-plug Efficiency (%)						
6	Useful Lifetime (hrs.)						
7	Footprint (m^2)						

*AD=Application Defined, **BPS=Beam Pointing Stability

A. CW Laser Checklist

Date	Project Name and #	Client Company		Rep Name:
		Name:		E-mail:
		Url:		Cell/Text #:

Laser Mf._____ Brand Name:_____ Model:_____

A	CW Parameter	Parameter Value (*AD)	Parameter Value (Actual)	PV Diff.	Relative Fluctuation Stability (%) (AD*)	Relative Fluctuation Stability (%) (Actual)	RFS Diff.
1	Power(W)						
2	*Wavelength (nm)*						
3	*Beam Width (mm)*						
4	*Beam Ellipticity*						
5	M^2						
6	*Gaussian Fit (%)*						
7	*BPS(µR/°C)*						
8	*Polarization Ratio*						
C	Tacit Parameter	Parameter Value (*AD)	Parameter Value (Actual)	PV Diff.	Relative Fluctuation Stability (%) (AD*)	Relative Fluctuation Stability (%) (Actual)	RFS Diff.
1	*Threshold*						
2	*Slope Efficiency (%)*						
3	*Warm-up Time (hrs.)*						
4	*Input AC (VAC)*						
5	*Wall-plug Efficiency (%)*						
6	*Useful Lifetime (hrs.)*						
7	*Footprint (m^2)*						

*AD=Application Defined, **BPS=Beam Pointing Stability

B. Pulse Laser Checklist

Date	Project Name and #	Client Company	Rep Name:	
		Name:	E-mail:	
		Url:	Cell/Text #:	

Laser Mf._____ Brand Name:_____ Model:_____

B	Pulse Parameter	Parameter Value (*AD)	Parameter Value (Actual)	PV Diff.	Relative Fluctuation Stability (%) (AD*)	Relative Fluctuation Stability (%) (Actual)	RFS Diff.
1	Average Power (W)						
2	Energy Per Pulse (mJ/Pulse)						
3	Pulse Duration						
4	Duty Cycle						
5	Rep Rate (kHz)						
6	Peak Power (W)						
C	**Tacit Parameter**	Parameter Value (*AD)	Parameter Value (Actual)	PV Diff.	Relative Fluctuation Stability (%) (AD*)	Relative Fluctuation Stability (%) (Actual)	RFS Diff.
1	Threshold						
2	Slope Efficiency (%)						
3	Warm-up Time (hrs.)						
4	Input AC (VAC)						
5	Wall-plug Efficiency (%)						
6	Useful Lifetime (hrs.)						
7	Footprint (m^2)						

*AD=Application Defined, **BPS=Beam Pointing Stability

Personal Notes

It's time to re-order your next batch of Laser Transactional/Performance Checklist booklets. Send an e-mail to orders@laserpronet.com for bulk discount purchases, or buy small amounts through amazon.

2. Self-Test

Laser Transactional/Performance Specifications a.k.a. Final Tests.

    A. CW Laser Transactional Specifications and Performance [1-55]
    B. Pulse Laser Transactional Specifications and Performance [56-77]
    C. Tacit Laser Performance Specifications [78-97]

A. CW Laser Transactional/Performance Specifications A.k.a. Final Tests

    CW 1. Power [1-5]
    CW 2. Power Stability [6-10]
    CW 3. Wavelength [11-15]
    CW 4. Beam Width and Ellipticity [16-20]
    CW 5. Beam Divergence [21-25]
    CW 6. M^2 [26-30]
    CW 7. Gaussian Fit [31-35]
    CW 8. Beam Pointing Stability [36-40]
    CW 9. Polarization Ratio [41-45]
    CW 10. Peak-to-peak Noise [46-50]
    CW 11. RMS Noise [51-55]

CW 1. Power [1-5]

1. Power is the rate at which_____exits the laser/gets deposited onto the target
    a. frequency
    b. heat
    c. energy
    d. all the above
    e. None of the above

2. Laser power is measured in
    a. Joules
    b. Watts
    c. Hertz
    d. None of the above

3. One Watt is one_____.
    a. Joule
    b. Joule/meter
    c. Joule/second
    d. Any of the above
    e. None of the above

4. During the laser warm-up period the laser would be in _____ operational mode(s).
    a. small signal gain
    b. saturation
    c. steady state
    d. a and b
    e. None of the above

5. Laser power is highest during the laser's_____phase of operation
    a. small signal
    b. saturation
    c. steady state
    d. Any of the above
    e. None of the above

CW 2. Power Stability [6-10]

6. A laser can fail power stability burn-in tests if the laser _____.
    a. power exceeds the user specification
    b. stops lasing
    c. a and b
    d. None of the above

7. Laser burn-in power stability tests take place during the _____operational mode(s) of the laser.
    a. small signal gain
    b. saturation
    c. steady state
    d. a and b
    e. None of the above

8. Laser power instability is suppressed in the_____Mode
    a. Current
    b. Power
    c. a and b
    d. None of the above

9. A laser will have the highest instability in the _____ Mode
    a. Current
    b. Power
    c. A and b
    d. None of the above

10. Two lasers that have the same power levels_____.
    a. must have the same power stability
    b. can have different stability levels
    c. any of the above
    d. none of the above

CW 3. Wavelength [11-15]

11. Monochromatic lasers are designed to output _____ wavelength.
    a. one
    b. more than one
    c. a or b
    d. None of the above

12. Laser wavelength(s) can be measured using a _____.
    a. spectrum analyzer
    b. spectrometer
    c. a and b
    d. None of the above

13. _____ nm is/are an invisible laser wavelength(s)
    a. 1064
    b. 355
    c. 266
    d. All the above
    e. None of the above

14. _____ nm wavelength(s) would require an IR viewer to be observed by the human eye.
    a. 1064
    b. 355
    c. 266
    d. All the above
    e. None of the above

15. _____ wavelength can optimally perform a specific application/job.
    a. A specific
    b. Any
    c. A and b
    d. None of the above

CW 4. Beam Width and Ellipticity [16 -20]

16. For a given laser resonator, the smallest beam spot size is only possible if it is composed of _____ mode(s).
    a. fundamental transverse
    b. TEM_{00}
    c. Higher order
    d. a and b
    e. None of the above

17. A Gaussian laser beam diameter/width is by default measured at _____ of maximum laser intensity.
    a. 50 %
    b. 13.5%
    c. $\frac{1}{e^2}$
    d. b and c
    e. None of the above

18. The beam waist size of a focused laser beam depends on
    a. laser wavelength
    b. focal length of the focusing/positive lens
    c. a and b
    d. None of the above

19. Beam ellipticity is also known as beam _____.
    a. Roundedness
    b. Circularity
    c. All of the above
    d. None of the above

20. Given a Gaussian beam whose transfer profile exhibits a D_y diameter of 100 microns (i.e. 100 x 10^{-6} mm) and a D_x diameter of 102 microns. The beam ellipticity is
    a. 1.02
    b. .98
    c. 1.0
    d. a and b
    e. none of the above

CW 5. Beam Divergence [21-25]

21. In commercial literature beam divergence is measured or characterized in_____.
    a. degrees
    b. radians
    c. a and b
    d. None of the above

22. Beam divergence can be specified as_____ angle.
    a. half
    b. full
    c. All the above
    d. None of the above

23. A_____ is needed to improve a laser beam's divergence
    a. Collimator
    b. Telescope
    c. A and b
    d. None of the above

24. A laser beam in the fundamental transverse mode has the minimum_____.
    a. angle of divergence achievable
    b. beam waist if focused
    c. a and b
    d. None of the above

25. An expanded collimated laser beam results in _____.
    a. low divergence
    b. smaller beam waist when/if focused
    c. a and b
    d. None of the above

CW 6. M^2 [26-30]

26. A truly TEM_{00} laser beam has a(n)
    a. M^2 value of 1
    b. Gaussian transverse profile
    c. a and b
    d. None of the above

27. A beam with a M^2 value of one ($M^2=1$) has the
    a. largest beam waist achievable
    b. largest angle of divergence achievable
    c. all the above
    d. None of the above

28. If a laser beam has a beam waist diameter of D_0 then the beam diameter/width at the either edge of the Rayleigh Range is given by
    a. $D_0\sqrt{2}$
    b. $2\sqrt{D_0}$
    c. $2D_0$
    d. None of the above

29. A TEM_{00} laser beam has the smallest possible beam _____.
    a. waist
    b. divergence.
    c. M^2 value
    d. All the above
    e. None of the above

30. The M^2 value of a laser beam can be used as an indicator of how well the beam will _____ if an appropriate lens is put in its path.
    a. focus
    b. diverge
    c. All the above
    d. None of the above

CW 7. Gaussian Fit [31-35]

31. The Gaussian Fit characterizes a laser beam's _____ mode(s).
    a. longitudinal
    b. transverse
    c. a or b
    d. None of the above

32. The Gaussian Fit's results are comparable to the _____ of a laser.
    a. M^2 value
    b. power
    c. a and b
    d. None of the above

33. A Gaussian Fit of 100% is an indication of a _____ beam
    a. TEM_{oo}
    b. multimode
    c. a and b
    d. None of the above

34. A Gaussian profile is the same as a(n) _____.
    a. M^2
    b. Beam Divergence
    c. Astigmatism
    d. All of the above
    e. None of the above

35. A Gaussian Fit is measured using a(n) _____.
    a. oscilloscope
    b. power meter
    c. energy meter
    d. beam profiler
    e. None of the above

CW 8. Beam Pointing Stability [36-40]

36. Beam Wander/Pointing Stability _____
    a. is sensitive temperature changes
    b. can be detected using a power meter
    c. is characterized as angular displacement per unit temperature
    d. a and c
    e. None of the above

37. Beam Pointing Stability is/can be measured as_____.
    a. Angular Displacement/Temperature.
    b. Linear Displacement/Temperature
    c. Temperature/ Angular Displacement
    d. Temperature
    e. None of the above

38. Beam Pointing Stability best results are achieved when _____ is/are placed on a vibration-free platform.
    a. laser
    b. profiler
    c. a and b
    d. None of the above

39. Beam Pointing Stability tests must be performed _____ to get accurate results
    a. on a vibration-free bench/platform.
    b. chilled chamber
    c. only when the laser output is fiber-coupled
    d. All the above
    e. None of the above

40. Beam Pointing Stability Tests are the same as _____ Stability Tests
    a. mechanical
    b. current
    c. voltage
    d. All the above
    e. None of the above

CW 9. Polarization Ratio [41-45]

41. Polarization Ratio is a measure of_____field in a laser beam.
    a. two orthogonal (right-angled) components of the electric
    b. two orthogonal (right-angled) components of the optical
    c. two orthogonal (right-angled) components of the magnetic
    d. a and b
    e. None of the above

42. Polarization Ratio is a ratio of_____values in a laser beam.
    a. Powers
    b. Energies
    c. Velocities
    d. a and b
    e. None of the above

43. Polarization Ratio of a laser beam has units of_____.
    a. Watts
    b. Joules
    c. Meters/second
    d. None of the above

44. If a laser beam is 100 % linearly polarized and is being passed through a linear polarizer that is being rotated, then a power meter receiving it would read a minimum value at _____degrees from the maximum power reading.
    a. 45
    b. 90
    c. 180
    d. None of the above

45. If a laser beam is 100 % linearly polarized and is being passed through a linear polarizer that is being rotated, then a power meter receiving it would read a minimum value of _____ Watts.
    a. 0
    b. 100
    c. 180
    d. Any of the above
    e. None of the above

CW 10. Peak-to-peak Noise [46-50]

46. When a laser beam signal is captured by sensor and transformed to voltage it can be split into_____ component(s)
    a. AC
    b. DC
    c. AC/DC
    d. a and b
    e. None of the above

47. A laser beam voltage that represents the "good" or desired laser beam is the_____voltage
    a. AC
    b. DC
    c. AC/DC
    d. All of the above
    e. None of the above

48. A laser beam voltage that represents the "bad" or undesired laser beam is the_____voltage
    a. AC
    b. DC
    c. AC/DC
    d. All the above
    e. None of the above

49. A voltage that represents the "bad" or undesired laser beam is directly measured from the laser beam's oscilloscope signal as_____peak-to-peak voltage
    a. AC
    b. DC
    c. AC/DC
    d. All the above
    e. None of the above

50. Peak-to-peak noise quantifies_____fluctuations over the "good" DC signal/voltage of a laser beam.
    a. AC
    b. DC
    c. AC/DC
    d. All the above
    e. None of the above

CW 11. RMS Noise [51-55]

51. The acronym RMS stands for
    a. Root-mean-square
    b. Radical-mean-square
    c. Root-mean-sum
    d. a and b
    e. None of the above

52. The RMS value of an electrical signal can be derived from _____ signals.
    a. DC
    b. AC
    c. AC/DC
    d. All the above
    e. None of the above

53. The RMS value of an electrical signal represents the _____ equivalent of an AC signal.
    a. DC
    b. polyphase
    c. AC/DC
    d. All the above
    e. None of the above

54. The voltage of a battery is a _____ signal.
    a. DC
    b. AC
    c. AC/DC
    d. All the above
    e. None of the above

55. RMS noise quantifies _____ fluctuations over the "good" DC signal/voltage of a laser beam.
    a. AC
    b. DC
    c. AC/DC
    d. All the above
    e. None of the above

B. Pulse Laser Transactional/Performance Specifications a.k.a. Final Tests

Pulse 1. Pulsed Power Output [56-60]
Pulse 2. Energy per pulse [61-65]
Pulse 3. Pulse Duration [66-70]
Pulse 4. Duty Cycle [71-75]
Pulse 5. Pulse Repetition Rate [76-77]

Pulsed 1. Pulsed Power Output [56-60]

56. Pulsed Peak Power is measured in
    a. Joules
    b. Watts
    c. Hertz
    d. All the above
    e. None of the above

57. Pulsed Average Power is measured in
    a. Joules
    b. Watts
    c. Hertz
    d. All the above
    e. None of the above

58. Pulsed Average Power is calculated as _____
    a. Energy Per pulse/Period
    b. Repetition Rate x Pulse Energy
    c. Peak Power x Duty Cycle
    d. All the above
    e. None of the above

59. Pulsed laser output is characterized by_____.
    a. Power
    b. Average Power
    c. Peak Power
    d. b and c
    e. None of the above

60. Average Power of a pulsed laser can be calculated as _____
    a. Energy Per pulse/Period
    b. Repetition Rate x Pulse Energy
    c. Energy Per Pulse/Pulse Duration
    d. a and b
    e. None of the above

Pulse 2. Energy Per Pulse [61-65]

61. Laser energy is measured in _____,
    a. Joules
    b. Watts
    c. Hertz
    d. All the above
    e. None of the above

62. Energy Per Pulse can be calculated as _____.
    a. Period x Average Power
    b. PRR x Average Power
    c. a and b
    d. All the above
    e. None of the above

63. There is/are _____ pulse(s) in a Period.
    a. one
    b. two
    c. one hundred
    d. any of the above
    e. None of the above

64. Energy Per pulse is an expression of a _____.
    a. quantity
    b. rate
    c. vector
    d. a and b
    e. None of the above

65. Energy Per pulse can be increased if _____ is decreased.
    a. PRR
    b. Period
    c. A and b
    d. None of the above
    e. All the above

Pulse 3. Pulse Duration [66-70]

66. Pulse Duration is also known as_____.
    a. Pulse Width
    b. Pulse Length
    c. FWHM
    d. a, b and c
    e. None of the above

67. Q-switch laser pulses are in the_____ second time regime
    a. Femto-
    b. nano
    c. milli
    d. b and c
    e. None of the above

68. Mode-locked laser shortest pulses are in the _____ second time regime.
    a. femto-
    b. nano-
    c. pico-
    d. milli-
    e. None of the above

69. The acronym FWHM stands for_____.
    a. Femto Width at Half Minimum
    b. Full width at Half Max
    c. Femto Width at Half Max
    d. b and c
    e. None of the above

70. Pulse Duration is measured in_____.
    a. Joules
    b. Watts
    c. Hertz
    d. All the above
    e. None of the above

Pulse 4. Duty Cycle [71-75]

71. Duty Cycle can be calculated as_____.
    a. Pulse Duration/Period.
    b. Pulse Length/FWHM
    c. FWHM/Period
    d. a and c
    e. None of the above

72. The Duty Cycle of a pulsed laser enumerates on_____.
    a. fraction of time a laser pulse is on in a period
    b. number of pulsed hitting a target in a period
    c. number of pulses output in a second.
    d. a and b
    e. None of the above

73. Altering the_____ of a laser will affect Duty Cycle
    a. Period
    b. FWHM
    c. pump energy
    d. a and b
    e. None of the above

74. A laser with a Duty Cycle of 100%_____.
    a. is Cw
    b. has a period equal to its FWHM
    c. must be mode-locked
    d. a and b
    e. None of the above

75. The smaller the Duty Cycle the_____ the Pulse Duration
    a. longer
    b. shorter
    c. Any of the above
    d. None of the above

Pulse 5. Pulse Repetition Rate (PRR) [76-77]

76. PRR is measured in_____.
    a. Joules
    b. Watts
    c. Hertz
    d. All the above
    e. None of the above

77. PRR is also known as_____'.
    a. Repetition Rate (Rep Rate)
    b. Pulse Repetition Frequency (PRF)
    c. All the above
    d. None of the above

C. Tacit Laser Performance Specifications [78-97]

    Tacit. 1 Relative Stability Fluctuations
    Tacit. 2 Threshold, Slope Efficiency, and Warm-up Time
    Tacit. 3 Power Supplies and Wall Plug Efficiency
    Tacit. 4 Useful Lifetime and Footprint

Tacit. 1 Relative Stability Fluctuations [78-82]

78. Laser beam performance parameter fluctuations general exhibit a_____distribution
    a. Normal
    b. Bell
    c. Gaussian
    d. All the above
    e. None of the above

79. _____level(s) of laser beam performance parameter fluctuation will be acceptable for any application.
    a. Some
    b. No
    c. All

80. The larger the standard deviation the_____precise a measurement/parameter value
    a. more
    b. less
    c. any of the above
    d. None of the above

81. Relative Stability Fluctuation discloses a specific parameter's performance_____in the laser output.
    a. precision
    b. average
    c. standard deviation
    d. all of the above
    e. None of the above

82. Relative Stability Fluctuations of laser performance parameters_____be disclosed to the user/buyer by a laser manufacturer unless otherwise requested
    a. will
    b. will not

Tacit. 2 Threshold, Slope Efficiency, and Warm-up Time [83-87]

83. Threshold is when a laser_____.
    a. starts to output
    b. when steady state has been reached
    c. a and b
    d. None of the above

84. After turning a laser on and before threshold a laser _____.
    a. does not consume any energy
    b. consumes energy
    c. does not output
    d. b and c
    e. None of the above

85. Warm-up time includes_____time(s).
    a. small signal gain
    b. saturation
    c. steady state
    d. a and b
    e. None of the above

86. Slope efficiency_____threshold.
    a. is a ratio of the output to input energy or power before
    b. reveals the fraction of energy input that emerges as laser output above
    c. a or b
    d. None of the above

87. Slope Efficiency is the same as_____Efficiency
    a. Conversion
    b. Electrical-to-optical
    c. Wall-plug
    d. All the above
    e. None of the above

Tacit.3 Electrical Outlets, Power Supplies, and Wall Plug Efficiency [88-92]

88. Electrical signals drawn by power supplies from electrical AC outlets convert the signals to_____.
    a. DC
    b. AC/DC
    c. Any of the above
    d. None of the above

89. Electrical signals drawn by power supplies from electrical AC outlets can_____.
    a. be of any AC Voltage
    b. only be 120 VAC
    c. only be 240 VAC
    d. a and b
    e. None of the above

90. Power Supply components include_____.
    a. Transformers
    b. Bridge rectifiers
    c. a and b
    d. None of the above

91. Lasers are run using_____electricity
    a. AC
    b. DC
    c. Any of the above
    d. None of the above

92. Wall-plug Efficiency is the ratio of_____.
    a. Input electrical power to output optical power
    b. Output optical power to input electrical power
    c. a or b
    d. None of the above

Tacit. 4 Useful Lifetime and Footprint [93-97]

93. Laser useful life-time information_____.
    a. is always 5, 000 hours
    b. can be predicted/estimated
    c. is generally, or may be, known by a laser manufacturer
    d. b and c
    e. None of the above

94. Laser lifetime tests are conducted using_____.
    a. Accelerated Aging Tests
    b. Pump Energy Aging Tests
    c. a and b
    d. None of the above

95. If a laser is to produce powers more than a kilowatt it must _____
    a. weight at least 10 pounds
    b. have a footprint at least 5 square feet
    c. All the above
    d. None of the above

96. Laser lifetime knowledge is useful to their users for _____.
    a. estimating the Return on Investment (ROI)
    b. forecasting and budgeting
    c. a and b
    d. None of the above

97. Laser useful life-time information is made available to the user/buyer _____.
    a. only upon request
    b. upon purchase
    c. through a court ruling
    d. None of the above

Take a graded version of this test online

www.freelaseriqscan.com

http://www.freelaseriqscan.com

Laser Test 1.3X Laser Transactional/Performance Specifications a.k.a. Final Tests

Order this course for your department or company

Laser Fundamentals and Performance Specifications

Course Description

If you are uncomfortable working with lasers as "black boxes" and would like to have a basic understanding of their inner workings, this introductory course will be of benefit to you. The workshop will cover the basic principles common to the operation of any laser/laser system. Next, we will discuss laser components and their functionality. Components covered will include laser pumps/energy sources, mirrors, active media, nonlinear crystals, and Q-switches. The properties of laser beams will be described in terms of some of their common performance specifications such as longitudinal modes and monochromaticity, transverse electromagnetic (TEM) modes and focusability, continuous wave (CW) power, peak power and power stability. Laser slope and wall-plug efficiencies will also be discussed.

Learning Outcomes

This course will enable you to:

- describe the overall inner workings of any laser
- describe the functionality of the key laser components
- know the difference between how acousto- and electro-optic Q-switches work
- explain how each key component in a laser may contribute to laser performance
- intelligently engage your clients or customers using proper laser terminology
- build stronger relationships with clients and customers by demonstrating product knowledge

- obtain the technical knowledge and confidence to enhance your job performance and rise above the competition, inside and outside your company

Intended Audience

Managers, engineers, technicians, assemblers, sales/marketing, customer service, and other support staff. This short course will help cultivate a common/standardized understanding of lasers across the company.

About the Instructor

Sydney Sukuta is currently a Laser Technology professor in Silicon Valley. He also has industry experience working for some of the world's leading laser manufacturers in Silicon Valley where he saw first-hand the issues they encounter on a daily basis. In response, Dr. Sukuta developed prescriptive short courses to help absolve most of these issues.

Order your course today for on-site, online or as part of a resort retreat.

orders@laserpronet.com

Laser Tech Book Catalog

#	ISBN	Title	# of Copies Needed
1	ISBN-10: 1724808303 ISBN-13: 978-1724808301	Basic Laser Technology: Comprehensive Course Notes and Workbook	
2	ISBN-10: 1725881314 ISBN-13: 978-1725881310	Solid State Lasers and their Common Problems: Comprehensive Courses Notes and Workbook	
3	ISBN-10: 1796522872 ISBN-13: 978-1796522877	Laser Advanced Concepts, and Common Problems and Practical Solutions: Comprehensive Course Notes and Workbook	
4	ISBN-10: 1727678699 ISBN-13: 978 1727678697	Good Lab and Presentation Practices: Lab Notebook with Tips on How to Protect Intellectual Property (IP)	
5	ISBN-10: 1721540539 ISBN-13: 978-1721540532	Good Laser Lab and Manufacturing Practices (GLLMP): Laser Lab Fundamentals and Performance Tests	
6	ISBN-10: 1725880385 ISBN-13: 978-1725880382	Good Laser Lab and Manufacturing Practices (GLLMPs): Laser Fabrication, Factory-level Tuning and Performance Tests	
7	SBN-10: 1796517658 ISBN-13: 978-1796517651	Good Laser Lab and Manufacturing Practices (GLLMPs): Laser Lab Instrumentation and Opto-Electronic Analyzers	
8	ISBN-10: 1541194543 ISBN-13: 978-1541194540	Laser Transactional/Performance Specifications: General Checklist and Self-Test	

orders@laserpronet.com

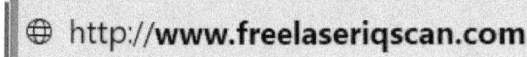

Log in directly

OR

Log in via our website

FREE Laser IQ Scan/Test
Percentile-ranked test results out and e-mailed instantly.
Only the best will deliver the best products and services.

Experienced *Professor with a demonstrated history of working with industry to produce employable and productive graduates.*

www.ingramcontent.com/pod-product-compliance
Lightning Source LLC
Chambersburg PA
CBHW061206180526
45170CB00002B/977